Marvellous Maths
Multiplication

By David L. Stienecker

Illustrations by Richard Maccabe

CHERRYTREE BOOKS

A Cherrytree Book

Adapted by A S Publishing from
Discovering Math
First published by Benchmark Books
Copyright © Marshall Cavendish Corporation 1996
Series created by Blackbirch Graphics Inc
Series editor: Tanya Lee Stone

This edition first published in 1997
by Cherrytree Press Ltd
a subsidiary of
The Chivers Company Ltd
Windsor Bridge Road
Bath BA2 3AX

Copyright © Cherrytree Press Ltd 1997

British Library Cataloguing in Publication Data
Stienecker, David L.
    Multiplication
    1. Multiplication—Juvenile literature 2. Multiplication—
Problems, exercises, etc.—Juvenile literature
    I. Title
    372.7'2

ISBN 0 7451 5313 5

Printed and bound in the USA

Contents

Mr. Ferris's Big Wheel

Ferris wheels were named after, you guessed it, Mr Ferris. His full name was George W. Gale Ferris. Here's what a typical Ferris wheel looks like.

• Most modern Ferris wheels have between 12 and 16 passenger cars. How many passenger cars does this Ferris wheel have?

• How many people can each passenger car hold? To find out, just count the number of people in one of the cars in the picture.

• How many people can ride this Ferris wheel at one time? How can you find out?

Now back to Mr Ferris. He got his name attached to all those Ferris wheels because he came up with the idea. Then he built the biggest one. He built it for a special fair in the city of Chicago in America over a hundred years ago.

Mr Ferris's big wheel was over 80 metres high and weighed over 1,080 tonnes. The big wheel had 36 beautiful wooden passenger cars.

• It took Mr Ferris's big wheel 10 minutes to go round once. The standard trip was two complete turns. How many minutes did the standard trip take?

• Forty people could sit comfortably in each of the passenger cars. How many people could ride on Mr Ferris's big wheel at one time?

To find out how many, you can add or multiply. When there are several equal groups, it's usually easier to multiply.

• Sometimes as many as 60 people were stuffed into each car. How many people could ride the big wheel then?

Mr Ferris's wonderful wheel stood in Chicago for ten years. Then it was taken apart and the pieces were sold as scrap!

One Hundred Facts!

This table shows one hundred multiplication facts! The factors are across the top and down the left side. The products are all the other numbers.

To find the product of 6 x 8, for example, look at the row that has 6 on the left. Look at the column that has 8 at the top. The product is the place where the 6 row crosses the 8 column: 6 x 8 = 48.

Look up words you don't understand on page 31.

X	0	1	2	3	4	5	6	7	8	9
0	0	0	0	0	0	0	0	0	0	0
1	0	1	2	3	4	5	6	7	8	9
2	0	2	4	6	8	10	12	14	16	18
3	0	3	6	9	12	15	18	21	24	27
4	0	4	8	12	16	20	24	28	32	36
5	0	5	10	15	20	25	30	35	40	45
6	0	6	12	18	24	30	36	42	48	54
7	0	7	14	21	28	35	42	49	56	63
8	0	8	16	24	32	40	48	56	64	72
9	0	9	18	27	36	45	54	63	72	81

Now have some fun with the table. Use it to find answers to some puzzles. First make copies of the table or lay tracing paper over it to work on.

To find the answer to this puzzle, shade in the product for each multiplication fact.

• A baker baked two dozen buns. All but eleven were eaten. How many were left?

2 x 2	2 x 3	3 x 3
4 x 3	5 x 3	6 x 3
2 x 5	2 x 6	3 x 6
4 x 6	5 x 6	6 x 6

X	0	1	2	3	4	5	6	7	8	9
0	0	0	0	0	0	0	0	0	0	0
1	0	1	2	3	4	5	6	7	8	9
2	0	2	4	6	8	10	12	14	16	18
3	0	3	6	9	12	15	18	21	24	27
4	0	4	8	12	16	20	24	28	32	36
5	0	5	10	15	20	25	30	35	40	45
6	0	6	12	18	24	30	36	42	48	54
7	0	7	14	21	28	35	42	49	56	63
8	0	8	16	24	32	40	48	56	64	72
9	0	9	18	27	36	45	54	63	72	81

When you've shaded in all the products, this is what the table looks like. The answer to the puzzle is eleven.

Here are some puzzles for you to try.

• If two is company and three is a crowd, what are four and five?

2 x 3	2 x 4	2 x 5	2 x 6	3 x 3
3 x 6	4 x 3	4 x 4	4 x 5	4 x 6
5 x 6	6 x 6	7 x 6	8 x 6	

• What is the beginning of everything?

3 x 2	3 x 3	3 x 4	3 x 5	4 x 2
5 x 2	5 x 3	5 x 4	6 x 2	7 x 2
7 x 3	7 x 4	7 x 5		

A dozen equals 12.

Make some designs of your own using multiplication facts. See if your friends can solve your puzzles.

Those Incredible 9's

One of the most incredible numbers is 9. If you don't believe it, just complete the 9's table on a separate piece of paper. Then add the digits of each product. What do you discover?

1 x 9 = 09   0 + 9 = ?

2 x 9 = 18   1 + 8 = ?

3 x 9 = ?

4 x 9 = ?

5 x 9 = ?

6 x 9 = ?

7 x 9 = ?

8 x 9 = ?

9 x 9 = ?

• Look at the products in the 9's table again. Count down the tens column. Then count up the ones column. What do you discover?

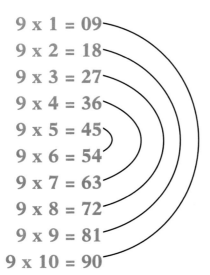

9 x 1 = 09

9 x 2 = 18

9 x 3 = 27

9 x 4 = 36

9 x 5 = 45

9 x 6 = 54

9 x 7 = 63

9 x 8 = 72

9 x 9 = 81

9 x 10 = 90

• Look closely at this version of the 9's table. Look at the products at both ends of the table going towards the middle. Do you notice anything unusual?

• You can discover some interesting number patterns using 9's. You may want to use a calculator. Otherwise, have plenty of paper ready because you'll be doing a lot of multiplication.

$$9 \times 9 + 7 = \ ?$$
$$98 \times 9 + 6 = \ ?$$
$$987 \times 9 + 5 = \ ?$$
$$9876 \times 9 + 4 = \ ?$$
$$98765 \times 9 + 3 = \ ?$$
$$987654 \times 9 + 2 = \ ?$$
$$9876543 \times 9 + 1 = \ ?$$
$$98765432 \times 9 + 0 = \ ?$$

Here's another pattern that you can make with 9's:

$$999999 \times 2 = \ ?$$
$$999999 \times 3 = \ ?$$
$$999999 \times 4 = \ ?$$
$$999999 \times 5 = \ ?$$
$$999999 \times 6 = \ ?$$
$$999999 \times 7 = \ ?$$
$$999999 \times 8 = \ ?$$
$$999999 \times 9 = \ ?$$

And another:

$$9 \times 1 + 2 = \ ?$$
$$9 \times 12 + 3 = \ ?$$
$$9 \times 123 + 4 = \ ?$$
$$9 \times 1234 + 5 = \ ?$$
$$9 \times 12345 + 6 = \ ?$$
$$9 \times 123456 + 7 = \ ?$$
$$9 \times 1234567 + 8 = \ ?$$
$$9 \times 12345678 + 9 = \ ?$$
$$9 \times 123456789 + 10 = \ ?$$

9

Calendar Challenge

Here's a challenge that you can win every time. You need a calendar, calculators, a pencil and paper.

Challenge your friends to a maths contest. Tell them that they can use calculators. You will use a paper and pencil.

Hand your friends the calendar and ask each of them to turn to a favourite month. Then tell them to draw a 3 x 3 box around any nine numbers like this.

Explain that the first person to find the sum of all nine numbers is the winner. Then start. With a little help from multiplication, you'll have the answer in no time. This is what you'll be doing while your friends are pushing buttons:

Jot down the smallest number in the box: **2**

Then add 8: **+8**

10

Multiply by 9: **x9**

It works every time! **90**

Calculator Words

Did you know your calculator can make words? If you punch in the right numbers, then turn your calculator upside down, you will find you've got a word!

Try this. Punch in 607. Turn your calculator upside down and you've got LOG. Try 376608 and see the word BOGGLE.

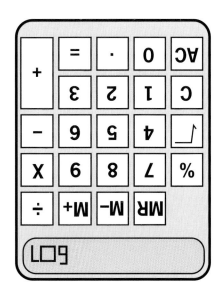

Complete these maths problems to find the words that go with the sentences.

Giddy as one of these.	17503 x 2
A pig by another name.	50 x 12 + 4
The earth is one of these.	3173 x 12
Travel on this over snow.	3691 x 125
What birds hatch from.	195 x 29 + 8
Eat too fast.	1021 x 371 + 15

• Try making some words of your own. Here are the letters that appear when the numbers are turned upside down:

1 = I	3 = E	4 = h	5 = S	6 = g
7 = L	8 = B	9 = b	0 = O	

11

Factor Trees

Have you ever heard of a factor tree? A factor tree is like a real tree. It has branches. The branches of a factor tree are made up of factors.

Here's an example of how to 'grow' a factor tree.

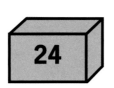

Begin by 'planting' a product. This product is 24.

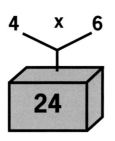

Then decide on two factors for your product. They will become the first row of branches in your factor tree.

The first two factors for this tree are 4 and 6. But you could have used 3 and 8 or 2 and 12.

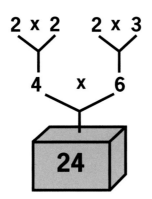

Now find factors for the first row of numbers in your factor tree. In this example, 2 and 2 are the factors for 4, and 2 and 3 are the factors for 6. These factors make the second row of branches in your tree.

Keep finding factors until the only factors left are one. Then your factor tree is complete. All the factors in the last row of your tree should equal the product you began with, in this case 24:

$$2 \times 2 = 4 \qquad 4 \times 2 = 8 \qquad 8 \times 3 = 24$$

See, it works!

Try your hand at 'growing' factor trees. Draw these trees on a sheet of paper. Fill in the missing factors.

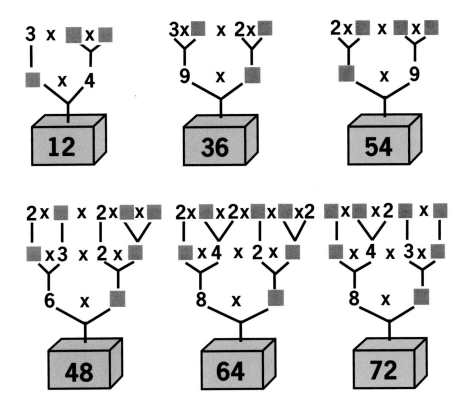

• Make some factor trees of your own. You can use any numbers for products. See how big a factor tree you can make.

Largest Product Game

Here's a game you can play with almost any number of people. These are the things you will need:

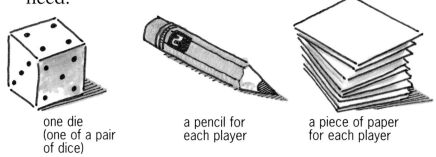

one die
(one of a pair
of dice)

a pencil for
each player

a piece of paper
for each player

Before playing, each player should draw a game board like this one on a sheet of paper:

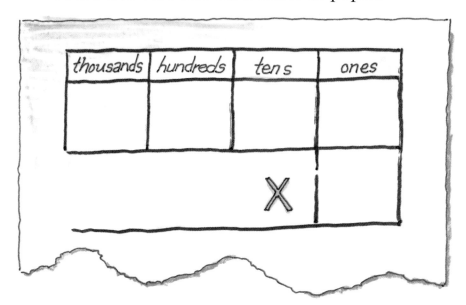

thousands	hundreds	tens	ones
		X	

How to play:

1. In turn, each player throws the die and reads the number aloud.

2. Each player writes the number in any of the boxes on his or her game board. Here's how one player's game board might look after a 3 and a 6 have been thrown.

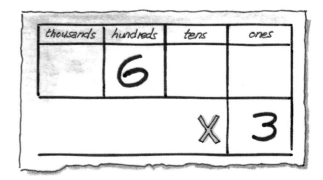

3. After five throws have been made, each player's game board will show a four-digit number multiplied by a one-digit number. This is how our player's game board looks at this stage of the game.

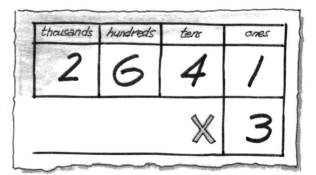

4. Each player should multiply the numbers on his or her game board. The winner is the person with the largest product.

• What is the product of our player's game board? How could the digits be arranged differently to make a larger product? Give it a try.

• Try this variation on the game. After everyone has multiplied their numbers, give them 60 seconds to rearrange the digits to make a larger product.

• Here's another way to play the game. Get players to multiply by a two-digit number instead of a one-digit number. Throw the die six times instead of five.

15

Area Puzzles

Find the area of a rectangle by multiplying the length by the width.

Four shapes like this one can be put together to make a square. Make four copies and give it a try. When you've finished, use a ruler to measure the sides, and multiply two of them to find the area of the square.

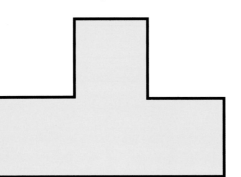

Now do the same with each of these figures.

A square is a rectangle with equal sides.

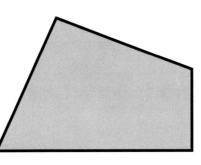

Use this shape four times to make a rectangle. Then find the area of the rectangle you made.

Digit Magic

Here's a multiplication trick to amaze your friends.

• Jot down this 'magic' number on a sheet of paper.

If you look carefully at the magic number, you will see it is easy to remember. It consists of all the digits written in order with the number 8 left out.

• Now ask a friend to tell you his or her favourite digit.

• Whatever number your friend chooses, multiply it in your head by 9. Write the product beneath the magic number. Suppose your friend chose the number 6. Think, '6 x 9 = 54'. Write the number 54 below the magic number like this:

• Then ask your friend to multiply the two numbers. To everyone's amazement, the product will be your friend's favourite digit. (You can use a calculator.)

Remember, the trick works with any digit. Try it several times just to convince yourself.

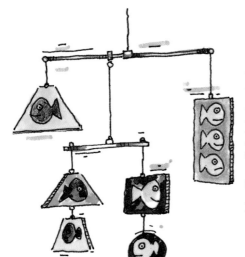

Mobile Challenge

Mobiles are movable sculptures. In fact, the word mobile means 'movable'. The parts of a mobile are put together so they will move in the air.

To make the arms of a mobile balance, you need to use a little multiplication. It's really very simple. The weight times the length of the rod on one side of the hanging wire must equal the weight times the length of the rod on the other side, like this.

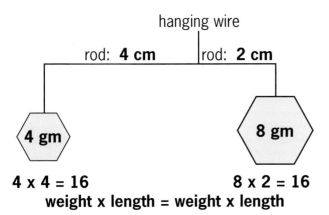

hanging wire

rod: **4 cm**   rod: **2 cm**

4 gm   **8 gm**

4 x 4 = 16   8 x 2 = 16

weight x length = weight x length

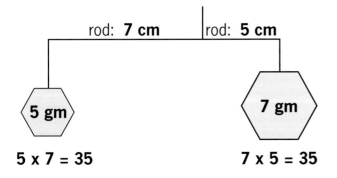

rod: **7 cm**   rod: **5 cm**

5 gm   **7 gm**

5 x 7 = 35   7 x 5 = 35

How many grams are hanging from the mobile above? Here's another way you could have made the mobile with the same total weight of 12 grams.

18

Here are two other ways to make the mobile with 12 grams hanging from it. Find the missing factors.

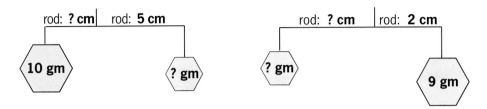

Here's something a little more challenging. Hint: start on the right side of the mobile.

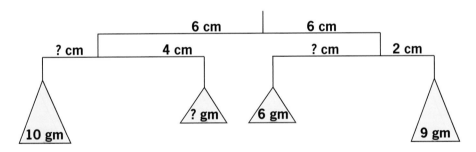

Here's something a *lot* more challenging.

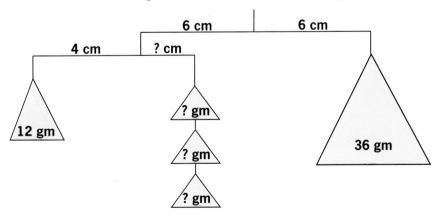

A Multiplication Code

The answers to the riddles below can be found by using the decoder. But first you have to decode the decoder. That's easy. Just find the product for each letter of the alphabet. Then solve the riddles.

DECODER

12	21	32	44	56	62	73	80	91
x3	x6	x3	x4	x2	x4	x6	x9	x7
36	?	?	?	?	?	?	?	?
A	**B**	**C**	**D**	**E**	**F**	**G**	**H**	**I**

123	213	124	327	503	416	524	718
x 2	x 3	x 4	x 2	x 5	x 3	x 4	x 2
?	?	?	?	?	?	?	?
J	**K**	**L**	**M**	**N**	**O**	**P**	**Q**

15	21	34	46	57	63	72	80	93
x11	x12	x23	x31	x30	x26	x14	x32	x15
?	?	?	?	?	?	?	?	?
R	**S**	**T**	**U**	**V**	**W**	**X**	**Y**	**Z**

What are six ducks in a box?

36   126-1248-1008   1248-248

1436-1426-36-96-639-112-165-252

Why is it bad manners to whisper?

126-112-96-36-1426-252-112   637-782   637-252

2515-1248-782   36-496-1248-1426-176

When is a car not a car?

1638-720-112-2515   637-782   637-252

782-1426-165-2515-637-2515-438

637-2515-782-1248   36

176-165-637-1710-112-1638-36-2560

Why does a hummingbird hum?

126-112-96-36-1426-252-112   637-782

176-1248-112-252   2515-1248-782

639-2515-1248-1638   782-720-112

1638-1248-165-176-252

Why should you do maths problems with a pencil?

126-112-96-36-1426-252-112   782-720-112

2096-112-2515-96-637-496

96-36-2515-2515-1248-782

176-1248   782-720-112-654

1638-637-782-720-1248-1426-782

2560-1248-1426

• Use the decoder to write messages of your own. Make up your own multiplication code.

21

Multiplication Maze

You can get through this maze quickly and easily with a little multiplication.

1. Start where it says IN.

2. When you come to a multiplication problem, find the product.

3. Look for the path with the correct answer. That's the way to go.

4. Continue solving problems and following paths until you come out of the maze.

Warning: Some of the multiplication problems don't have a correct answer on a path. If you reach one of these, you know you are in trouble. Time to backtrack. Good Luck!

Don't draw on the maze.

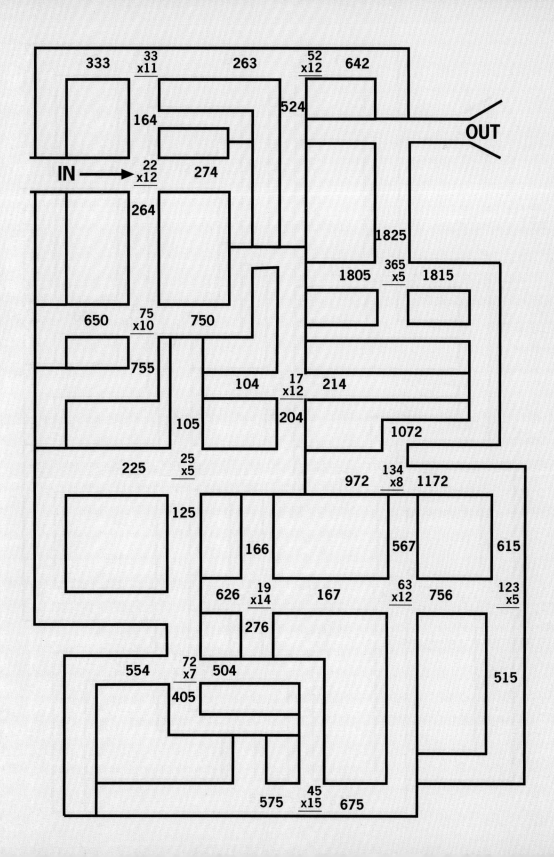

Fast 11's

You won't believe how fast you can multiply any number by 11 with this method. Here's how to do it.

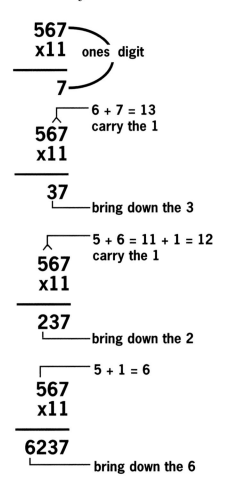

The ones digit of the answer is the same as the ones digit of the number you're multiplying. Just bring it down.

Now for the tens digit. Look at the tens digit in the number you are multiplying. Add it to the number on the right: 6 + 7 = 13. Carry as in ordinary addition.

Continue in the same way no matter how big the number is you are multiplying. Don't forget to add the number you carry.

The farthest left-hand digit in the answer is the left-hand digit in the number plus anything you carried.

Of course, you don't always have to carry. Try to find the products of these problems using this method.

137	246	345	538	623	754
x11	x11	x11	x11	x11	x11
?	?	?	?	?	?

Number Puzzle

You won't believe your eyes when you try this with your calculator.

1. Key in 37037.

2. Choose a number from 1 to 9. In your head multiply the number by 3.

3. Multiply 37037 by that number. For example, if you chose 3, you need to multiply by 9. If you chose 8, you need to multiply by 24.

The product will be a row of the number you chose!

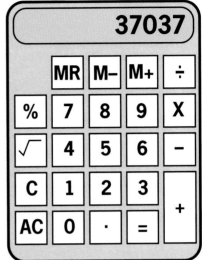

Find the Missing Digits

Here are some puzzles to challenge your grey matter. Find the missing digits in this problem.

Here's one of the simpler solutions. Try to find the other three.

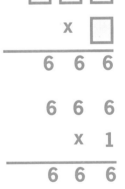

What are the missing digits in these two problems? There are four solutions for each one. How many can you find?

Make a Deal

Here's a deal you can make with a grown-up and come out a big winner!

Think of some chore you can do every day like washing dishes, taking out the rubbish or some other ghastly job. Then make this deal.

You will charge 1 penny the first day, and each day you'll charge twice as much as the day before. Who could turn down a deal like that? You may not believe it now, but you're going to get rich!

DAY	PAY
1	1p
2	2p
3	4p
4	8p
5	16p
6	32p
7	64p
8	£1.28
9	?
10	?

To see how rich, use a calculator to work out your pay for each day. Use a chart like this to keep track. Figure out how much you'll make in two weeks. Then figure out how much you'll make in a month.

Answers

Pages 4/5 Mr Ferris's Big Wheel

According to the picture, the Ferris wheel has 16 passenger cars. Each one can hold 3 people.

You can think of the number of people the Ferris wheel can hold as 16 groups of 3. To find out how many people can ride the Ferris wheel at one time, you can add 3 – 16 times. Or, you can multiply 16 by 3:

$$\begin{array}{r} 16 \\ \times 3 \\ \hline 48 \text{ people} \end{array}$$

If it takes Mr Ferris's big wheel 10 minutes to go round once, you can find out how long 2 complete turns will take by either adding or multiplying:

10 + 10 = 20 minutes

10 x 2 = 20 minutes

If 40 people are in each of 36 passenger cars, you can find out how many in all by multiplying this way:

$$\begin{array}{r} 36 \\ \times 40 \\ \hline 1440 \text{ people} \end{array}$$

Or this way:

$$\begin{array}{r} 36 \\ \times 10 \\ \hline 360 \end{array} \qquad \begin{array}{r} 360 \\ \times 4 \\ \hline 1440 \text{ people} \end{array}$$

If there are 60 people in each passenger car:

$$\begin{array}{r} 36 \\ \times 60 \\ \hline 2160 \text{ people} \end{array}$$

Or:

$$\begin{array}{r} 36 \\ \times 10 \\ \hline 360 \end{array} \qquad \begin{array}{r} 360 \\ \times 6 \\ \hline 2160 \end{array}$$

6/7 One Hundred Facts!

If two is company and three is a crowd, what are four and five?

X	0	1	2	3	4	5	6	7	8	9
0	0	0	0	0	0	0	0	0	0	0
1	0	1	2	3	4	5	6	7	8	9
2	0	2	4	6	8	10	12	14	16	18
3	0	3	6	9	12	15	18	21	24	27
4	0	4	8	12	16	20	24	28	32	36
5	0	5	10	15	20	25	30	35	40	45
6	0	6	12	18	24	30	36	42	48	54
7	0	7	14	21	28	35	42	49	56	63
8	0	8	16	24	32	40	48	56	64	72
9	0	9	18	27	36	45	54	63	72	81

What is the beginning of everything?

X	0	1	2	3	4	5	6	7	8	9
0	0	0	0	0	0	0	0	0	0	0
1	0	1	2	3	4	5	6	7	8	9
2	0	2	4	6	8	10	12	14	16	18
3	0	3	6	9	12	15	18	21	24	27
4	0	4	8	12	16	20	24	28	32	36
5	0	5	10	15	20	25	30	35	40	45
6	0	6	12	18	24	30	36	42	48	54
7	0	7	14	21	28	35	42	49	56	63
8	0	8	16	24	32	40	48	56	64	72
9	0	9	18	27	36	45	54	63	72	81

Answers

8/9 Those Incredible 9's

If you add the digits of each product, you discover that each sum equals 9:
$0 + 9 = 9$; $1 + 8 = 9$; $2 + 7 = 9$; $3 + 6 = 9$;
$4 + 5 = 9$; $5 + 4 = 9$; $6 + 3 = 9$; $7 + 2 = 9$;
$8 + 1 = 9$.

The tens column counts down by ones. The ones column counts up by ones.

The products at both ends of the 9's table, as you go towards the middle, are the same numbers turned round.

These are the number patterns that you can make:

$$9 \times 9 + 7 = 88$$
$$98 \times 9 + 6 = 888$$
$$987 \times 9 + 5 = 8888$$
$$9876 \times 9 + 4 = 88888$$
$$98765 \times 9 + 3 = 888888$$
$$987654 \times 9 + 2 = 8888888$$
$$9876543 \times 9 + 1 = 88888888$$
$$98765432 \times 9 + 0 = 888888888$$

$$999999 \times 2 = 1999998$$
$$999999 \times 3 = 2999997$$
$$999999 \times 4 = 3999996$$
$$999999 \times 5 = 4999995$$
$$999999 \times 6 = 5999994$$
$$999999 \times 7 = 6999993$$
$$999999 \times 8 = 7999992$$
$$999999 \times 9 = 8999991$$

$$9 \times 1 + 2 = 11$$
$$9 \times 12 + 3 = 111$$
$$9 \times 123 + 4 = 1111$$
$$9 \times 1234 + 5 = 11111$$
$$9 \times 12345 + 6 = 111111$$
$$9 \times 123456 + 7 = 1111111$$
$$9 \times 1234567 + 8 = 11111111$$
$$9 \times 12345678 + 9 = 111111111$$
$$9 \times 123456789 + 10 = 1111111111$$

10 Calendar Challenge

No answers.

11 Calculator Words

Giddy as one of these.
$17503 \times 2 = 35006 = $ goose

A pig by another name.
$50 \times 12 + 4 = 604 = $ hog

The earth is one of these.
$3173 \times 12 = 38076 = $ globe

Travel on this over snow.
$3691 \times 125 = 461375 = $ sleigh

What birds hatch from.
$195 \times 29 + 8 = 5663 = $ eggs

Eat too fast.
$1021 \times 371 + 15 = 378806 = $ gobble

12/13 Factor Trees

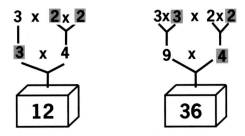

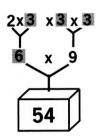

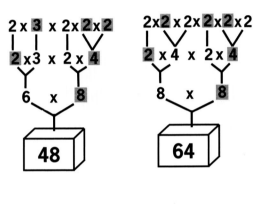

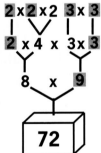

14/15 Largest Product Game

The product on the game board is 7,923.

The digits on the sample game board could be arranged in other ways to make larger products. Here are two:

6421 x 3 = 19,263
4621 x 3 = 13,863

16 Area Puzzles

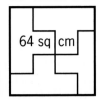

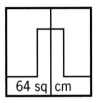

17 Digit Magic

No answers.

18/19 Mobile Challenge

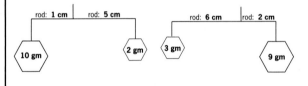

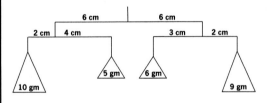

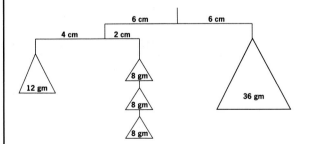

Answers

20/21 Multiplication Code

12	21	32	44	56	62	73	80	91
x3	x6	x3	x4	x2	x4	x6	x9	x7
36	126	96	176	112	248	438	720	637
A	**B**	**C**	**D**	**E**	**F**	**G**	**H**	**I**

123	213	124	327	503	416	524	718
x 2	x 3	x 4	x 2	x 5	x 3	x 4	x 2
246	639	496	654	2515	1248	2096	1436
J	**K**	**L**	**M**	**N**	**O**	**P**	**Q**

15	21	34	46	57	63	72	80	93
x11	x12	x23	x31	x30	x26	x14	x32	x15
165	252	782	1426	1710	1638	1008	2560	1395
R	**S**	**T**	**U**	**V**	**W**	**X**	**Y**	**Z**

What are six ducks in a box?
A box of quackers.

Why is it bad manners to whisper?
Because it is not aloud.

When is a car not a car?
When it is turning into a driveway.

Why does a hummingbird hum?
Because it does not know the words.

Why should you do maths problems with a pencil?
Because the pencil cannot do them without you.

22/23 Multiplication Maze
No answers.

24 Fast 11's

137	246	345
x11	x11	x11
1507	2706	3795

538	623	754
x11	x11	x11
5918	6853	8294

25 Number Puzzle
No answers.

Find the Missing Digits

111	333	222
x6	x2	x3
666	666	666

444	222	148	111
x1	x2	x3	x4
444	444	444	444

888	444	296	111
x1	x2	x3	x8
888	888	888	888

26 Make a Deal
At the end of two weeks (14 days) you will receive £81.92.

At the end of one month (30 days) you will receive:

£5,368,708.80

Glossary

area The number of square units needed to cover a surface.
Example: The area of this rectangle is 6 square centimetres.

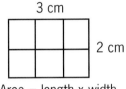

3 cm

2 cm

Area = length x width

column An up and down arrangement of things.

digits The symbols used to write numerals: 0, 1, 2, 3, 4, 5, 6, 7, 8 and 9.

factors Numbers that are multiplied together.

$$5 \times 7 = 35$$
factors

multiplication A mathematical operation with two numbers that results in a product.

multiplication fact A statement such as 6 x 8 = 48 is a multiplication fact.

product The answer to a multiplication problem.

$$5 \times 7 = 35$$
product

rectangle A shape with four sides and four right angles.
Example:

row A number of objects arranged in a straight line.

square A rectangle with all four sides straight and the same length.

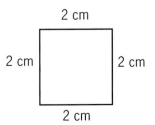

2 cm

2 cm   2 cm

2 cm

31

Index

A
area, 16

C
calculator, 9, 10, 11, 17, 25, 26
calculator words, 11
column, 6, 8

D
decoder, 20, 21
die, 14, 15
digits, 8, 15, 17, 24, 25

F
factor trees, 12–13
factors, 6, 12, 13, 19

L
length, 16, 18

M
multiplication codes, 20–21

multiplication facts, 6, 7
multiplication maze, 22–23

N
number patterns, 9

P
product, 6, 7, 8, 12, 13, 15, 17, 20, 22, 25

R
rectangle, 16
row, 6, 12, 13, 25

S
square, 16
sum, 10

W
width, 16